ENERGY SECTOR STANDARD
OF THE PEOPLE' S REPUBLIC OF CHINA

中华人民共和国能源行业标准

# Code for Design of DC Generation System for Photovoltaic Power Stations

# 光伏发电站直流发电系统设计规范

**NB/T 32045-2018**

Chief Development Department: China Renewable Energy Engineering Institute

Approval Department: National Energy Administration of the People's Republic of China

Implementation Date: October 1, 2018

China Water & Power Press

中国水利水电出版社

Beijing 2024

图书在版编目（CIP）数据
光伏发电站直流发电系统设计规范 : NB/T 32045-2018 = Code for Design of DC Generation System for Photovoltaic Power Stations（NB/T 32045-2018） : 英文 / 国家能源局发布. -- 北京 : 中国水利水电出版社, 2024. 6. -- ISBN 978-7-5226-2664-2
Ⅰ. TM615-65
中国国家版本馆CIP数据核字第2024DG3063号

ENERGY SECTOR STANDARD
OF THE PEOPLE'S REPUBLIC OF CHINA
中华人民共和国能源行业标准

**Code for Design of DC Generation System for Photovoltaic Power Stations**

**光伏发电站直流发电系统设计规范**

**NB/T 32045-2018**

（英文版）

Issued by National Energy Administration of the People's Republic of China
国家能源局　发布
Translation organized by China Renewable Energy Engineering Institute
水电水利规划设计总院　组织翻译
Published by China Water & Power Press
中国水利水电出版社　出版发行
Tel: (+ 86 10) 68545888　68545874
sales@mwr.gov.cn
Account name: China Water & Power Press
Address: No.1, Yuyuantan Nanlu, Haidian District, Beijing 100038, China
http: //www.waterpub.com.cn
中国水利水电出版社微机排版中心　排版
北京中献拓方科技发展有限公司　印刷
184mm×260mm　16 开本　1.75 印张　55 千字
2024 年 6 月第 1 版　2024 年 6 月第 1 次印刷
**Price（定价）: ¥ 280.00**

# Introduction

This English version is one of China's energy sector standard series in English. Its translation was organized by China Renewable Energy Engineering Institute authorized by National Energy Administration of the People's Republic of China in compliance with relevant procedures and stipulations. This English version was issued by National Energy Administration of the People's Republic of China in Announcement [2023] No. 8 dated December 28, 2023.

This version was translated from the Chinese Standard NB/T 32045-2018, *Code for Design of DC Generation System for Photovoltaic Power Stations*, published by China Water & Power Press. The copyright is reserved by National Energy Administration of the People's Republic of China. In the event of any discrepancy in the implementation, the Chinese version shall prevail.

Many thanks go to the staff from the relevant standard development organizations and those who have provided generous assistance in the translation and review process.

For further improvement of the English version, any comments and suggestions are welcome and should be addressed to:

China Renewable Energy Engineering Institute
No. 2 Beixiaojie, Liupukang, Xicheng District, Beijing 100120, China
Website: www.creei.cn

Translating organization:

Shanghai Electric Power Design Institute Co., Ltd.

Translating staff:

| | | | |
|---|---|---|---|
| LIU Yulei | GONG Chunjing | ZHANG Rongguo | LIU Daqian |
| DENG Yu | WANG Qiangjie | HUANG Pengcheng | TANG Liang |
| YAN Jiayin | YANG Chen | | |

Review panel members:

| | |
|---|---|
| QIE Chunsheng | Senior English Translator |
| YAN Wenjun | Army Academy of Armored Forces, PLA |
| ZHANG Ming | Tsinghua University |
| GUO Jie | POWERCHINA Beijing Engineering Corporation Limited |

| | |
|---|---|
| LI Kejia | POWERCHINA Northwest Engineering Corporation Limited |
| SU Xin | Linyi University |
| NIE Ming | Shanghai Electric Power Design Institute Co., Ltd. |
| YUE Lei | China Renewable Energy Engineering Institute |

National Energy Administration of the People's Republic of China

# 翻译出版说明

本译本为国家能源局委托水电水利规划设计总院按照有关程序和规定，统一组织翻译的能源行业标准英文版系列译本之一。2023 年 12 月 28 日，国家能源局以 2023 年第 8 号公告予以公布。

本译本是根据中国水利水电出版社出版的《光伏发电站直流发电系统设计规范》NB/T 32045—2018 翻译的，著作权归国家能源局所有。在使用过程中，如出现异议，以中文版为准。

本译本在翻译和审核过程中，本标准编制单位及编制组有关成员给予了积极协助。

为不断提高本译本的质量，欢迎使用者提出意见和建议，并反馈给水电水利规划设计总院。

地址：北京市西城区六铺炕北小街 2 号
邮编：100120
网址：www.creei.cn

本译本翻译单位：上海电力设计院有限公司

本译本翻译人员：刘玉雷　龚春景　章荣国　刘大千
邓　宇　王强杰　黄鹏程　唐　良
严家茵　杨　晨

本译本审核人员：

郄春生　英语高级翻译

闫文军　中国人民解放军陆军装甲兵学院

张　明　清华大学

郭　洁　中国电建集团北京勘测设计研究院有限公司

李可佳　中国电建集团西北勘测设计研究院有限公司

苏　鑫　山东临沂大学

聂　明　上海电力设计院有限公司

岳　蕾　水电水利规划设计总院

国家能源局

# Announcement of National Energy Administration of the People's Republic of China [2018] No. 8

According to the requirements of Document GNJKJ [2009] No. 52, "Notice on Releasing the Energy Sector Standardization Administration Regulations (tentative) and detailed implementation rules issued by National Energy Administration of the People's Republic of China", 87 sector standards such as *Wellbore Quality Control Requirements for Coalbed Methane Directional Well*, including 47 energy standards (NB) and 40 electric power standards (DL), are issued by National Energy Administration of the People's Republic of China after due review and approval.

Attachment: Directory of Sector Standards

National Energy Administration of the People's Republic of China

June 6, 2018

Attachment:

**Directory of Sector Standards**

| Serial number | Standard No. | Title | Replaced standard No. | Adopted international standard No. | Approval date | Implementation date |
|---|---|---|---|---|---|---|
| ... | | | | | | |
| 9 | NB/T 32045-2018 | Code for Design of DC Generation System for Photovoltaic Power Stations | | | 2018-06-06 | 2018-10-01 |
| ... | | | | | | |

# Foreword

According to the requirements of Document GNKJ [2015] No. 12 issued by National Energy Administration of the People's Republic of China, "Notice on Releasing the Development and Revision Plan of the Second Batch of Energy Sector Standards in 2014", and after extensive investigation and research, summarization of practical experience, consultation of relevant advanced Chinese and foreign standards, and wide solicitation of opinions, the drafting group has prepared this code.

The main technical contents of this code include: basic requirements, DC generation system, selection of equipment and cable, equipment layout and installation, overvoltage protection and earthing, and protection and monitoring.

National Energy Administration of the People's Republic of China is in charge of the administration of this code. China Renewable Energy Engineering Institute has proposed this code and is responsible for its routine management. China Renewable Energy Engineering Institute is responsible for the explanation of specific technical contents. Comments and suggestions in the implementation of this code should be addressed to:

China Renewable Energy Engineering Institute
No. 2 Beixiaojie, Liupukang, Xicheng District, Beijing, 100120, China.

Chief development organizations:

China Renewable Energy Engineering Institute

Shanghai Electric Power Design Institute Co., Ltd.

Participating development organization:

POWERCHINA Northwest Engineering Corporation Limited

Chief drafting staff:

| | | | |
|---|---|---|---|
| GONG Chunjing | NIE Ming | ZHANG Rongguo | XIAO Bin |
| SHI Yaoyu | HUANG Qian | TIAN Lisha | HUANG Pengcheng |
| DENG Yu | ZHANG Ping | ZHOU Qin | GUO Zhenni |

Review panel members:

| | | | |
|---|---|---|---|
| LIU Guoyang | WEI Huixiao | WANG Biao | LIU Qigen |
| PENG Tiankui | WANG Junhua | HE Yibin | LI Xiupu |

SHU Zhenhuan CHEN Hao XIA Xiaodong ZHANG Xianli

DING Shipan ZHANG Shengzhong LU Jun ZHOU Qiulan

XUE Meijuan LI Shisheng

# Contents

# 1 General Provisions

**1.0.1** This code is formulated with a view to standardizing the design of DC generation system for photovoltaic power stations, to achieve the objectives of safety, reliability, economy, practicability, and technological advancement.

**1.0.2** This code is applicable to the construction, renovation and extension of grid-connected photovoltaic power stations.

**1.0.3** In addition to this code, the design of DC generation system for photovoltaic power stations shall comply with other current relevant standards of China.

# 2 Terms

**2.0.1** DC generation system of photovoltaic power stations

DC power generation and collection system from the photovoltaic (PV) modules to the inverter in a photovoltaic power station, including PV modules, DC combiner boxes, inverters and DC cables

**2.0.2** potential induced degradation (PID)

PV module performance degradation caused by the bias voltage between the circuits inside the PV module and their encapsulation materials, also known as voltage induced degradation

**2.0.3** DC/AC ratio

ratio between the sum of peak powers of the inverter-connected PV modules under standard test conditions and the rated output power of the inverter

# 3 Basic Requirements

**3.0.1** The design of DC generation system for photovoltaic power stations shall meet the requirements of photovoltaic power station planning, and utilize the land and solar energy resources in a rational, economical and effective way.

**3.0.2** The DC generation system of photovoltaic power stations shall meet the requirements of energy conservation, environmental protection, safety, etc.

**3.0.3** The DC generation system of the photovoltaic power station should prioritize the use of new technologies, new processes, novel equipment and novel materials after techno-economic comparison, if the safety and reliability requirements can be satisfied.

**3.0.4** All equipment and components of the DC generation system of photovoltaic power stations shall comply with the current relevant standards of China.

# 4 DC Generation System

## 4.1 General Requirements

**4.1.1** The system voltage of the DC generation system of photovoltaic power stations shall be determined through techno-economic comparison.

**4.1.2** The maximum open-circuit voltage of a PV string under local daytime extreme environmental conditions shall not exceed the system voltage of DC generation system of the photovoltaic power station.

**4.1.3** The DC/AC ratio between PV module and inverter in the DC generation system shall be determined through techno-economic comparison, considering the local solar energy resources, operating environment conditions, module installation mode, DC loss, etc.

**4.1.4** The electrical performance parameters of each PV module in the same string of a PV array shall be consistent, and the number of PV strings in series shall comply with the current national standard GB 50797, *Code for Design of Photovoltaic Power Station*.

**4.1.5** The working voltage, cable voltage drop, array orientation, PV panel tilt angle and shading of the PV strings connected to the same multi-channel maximum power point tracking (MPPT) module should be consistent.

**4.1.6** When the output characteristics of PV strings in an array differ greatly, the equipment with MPPT function should be selected.

**4.1.7** The working voltage of PV strings at the maximum power point should be within the MPPT voltage range of the inverter.

## 4.2 Electrical Wiring

**4.2.1** When centralized or distributed inverters are used, the photovoltaic power station should adopt a multi-level collection mode at the DC side. When string inverters are used, the DC-side collection function should be integrated into the inverter.

**4.2.2** Single-bus configuration should be adopted for the internal wiring of DC combiner boxes and DC distribution cabinets.

# 5 Selection of Equipment and Cable

## 5.1 PV Module

**5.1.1** PV modules shall be selected in terms of type, peak power, conversion efficiency, temperature coefficient, module size and weight, current-voltage curve, power-voltage curve, attenuation characteristics, etc.

**5.1.2** The performance parameters shall be verified by the operating environmental conditions including solar irradiance, ambient temperature, wind load and snow load.

**5.1.3** Anti-PID PV modules should be selected for high temperature, high humidity and high salt mist environment.

**5.1.4** The crystalline silicon PV modules or concentrated PV modules should be selected for areas with high solar irradiance and large direct normal irradiance. Thin-film PV modules should be selected for areas with low solar irradiance, large diffuse horizontal irradiance and high ambient temperature. The type of PV modules should be determined through techno-economic comparison.

**5.1.5** For building-mounted PV generation systems, PV modules selected should be in harmony with the buildings on which they are mounted. Building-integrated PV modules shall meet the technical requirements for building materials and structure members.

## 5.2 DC Combiner Box

**5.2.1** The DC combiner box shall be selected in terms of type, insulation level, voltage, temperature rise, ingress protection (IP) rating, number of input and output circuits, rated input and output currents, etc.

**5.2.2** The performance parameters shall be verified by the operating environment conditions including ambient temperature, relative humidity and altitude, taking into account the impact of operating environmental conditions such as pollution level and seismic intensity.

**5.2.3** The rated voltage of components on main electrical circuit of the DC combiner box shall not be lower than the voltage of the DC generation system.

**5.2.4** The incoming circuit of the DC combiner box shall be equipped with DC fuses or DC circuit breakers, and the outgoing side of the DC combiner box should be equipped with DC circuit breakers. DC fuses and DC circuit breakers shall be selected in accordance with the current national standards GB/T 13539.6, *Low-Voltage Fuses—Part 6: Supplementary Requirements for Fuse-*

*Links for the Protection of Solar Photovoltaic Energy Systems*; and GB/T 34581, *General Technical Requirements for Photovoltaic System DC Circuit Breakers*.

**5.2.5** Outdoor DC combiner boxes shall be corrosion- and rust-proof and protected from sunlight, and the IP rating shall not be inferior to IP54.

## 5.3 Inverter

**5.3.1** The selection of inverters shall comply with the current standards of China GB 50797, *Code for Design of Photovoltaic Power Station*; GB/T 30427, *Technical Requirements and Test Methods for Grid-Connected PV Inverters*; and NB/T 32004, *Technical Specification of PV Grid-Connected Inverter*.

**5.3.2** The unit capacity of inverters shall be selected according to the installed capacity of the photovoltaic power station through techno-economic comparison, considering DC loss, cable usage, civil engineering cost, module layout, etc.

**5.3.3** The impact on inverters of ambient temperature, relative humidity, altitude, seismic intensity and pollution level shall be taken into consideration.

**5.3.4** Anti-PID inverters should be selected for high temperature, high humidity and high salt mist environment.

**5.3.5** For inverters used in plateau areas with an altitude of 2000 m or above, plateau-specific products shall be selected or capacity-reduced measures be taken.

**5.3.6** Inverters with high weighted average conversion efficiency should be selected.

**5.3.7** For centralized and distributed inverters, DC circuit breakers should be provided at the DC input side.

**5.3.8** The rated voltage of components on main electrical circuit at DC side of the inverter shall not be lower than the design voltage of the DC generation system.

**5.3.9** DC insulation monitoring device should be provided at DC side of the inverter.

**5.3.10** The IP rating shall not be inferior to IP54 for outdoor inverters, should not be inferior to IP20 for indoor inverters.

## 5.4 DC Cable

**5.4.1** The selection of DC cables shall comply with the current standards of China GB 50217, *Code for Design of Cables of Electric Engineering*; and NB/T

42073, *Electric Cables for Photovoltaic Systems*.

**5.4.2** The selection of DC cables shall consider such factors as current-carrying capacity, thermal stability, correction coefficient, voltage drop, insulation withstand voltage and protection coordination.

**5.4.3** The connecting cables between PV strings shall be UV resistant.

**5.4.4** Twin-core cables should be used for electrical connection from DC combiner boxes to inverters.

**5.4.5** For centralized and distributed inverters, the maximum voltage drop at DC side from PV strings to DC combiner boxes and inverters should not exceed 2.0 % under standard test conditions, and the voltage drop of each string should be consistent.

**5.4.6** For string inverters, the maximum voltage drop at DC side from PV string to inverter should not exceed 1.0 % under standard test conditions.

**5.4.7** The cables shall be armored if they are directly buried, and may not be armored if laid in conduit.

# 6 Equipment Layout and Installation

## 6.1 PV Module Layout

**6.1.1** The characteristic curves of the PV modules in the same series circuit should be identical, and the deviation of peak working current should not exceed 0.1 A. The peak working voltage of PV modules in the same shunt circuit should be identical.

**6.1.2** The PV modules of the same current class shall be arranged in a centralized manner, and PV modules of higher current class should be arranged closer to the inverter.

**6.1.3** The horizontal or vertical layout of PV modules shall be determined through techno-economic comparison in terms of module specifications, site topography and support structure, etc.

**6.1.4** For building-mounted PV generation systems, the support structure should not cross the building deformation joint, and the PV modules shall be so arranged to avoid the pipes for roof smoke exhaust, air exhaust, and ventilation.

**6.1.5** The vertical distance from PV modules to building envelope shall satisfy the requirements of installation and ventilation for heat dissipation.

## 6.2 DC Combiner Box Layout

**6.2.1** DC combiner boxes should be arranged in the center of the collection area.

**6.2.2** DC combiner boxes should be mounted on PV support or independent support.

**6.2.3** DC combiner boxes shall be so arranged to facilitate repair and maintenance and be protected from direct sunlight.

**6.2.4** For the photovoltaic power station built on the water, DC combiner boxes shall be so arranged to facilitate operation and maintenance.

## 6.3 Inverter Layout

**6.3.1** Inverters should be arranged in the center of the PV array, and centralized or distributed inverters should be arranged close to the main passage.

**6.3.2** For the PV DC generation system built on the water, inverters should be arranged close to the shore or patrol passage if it is techno-economically rational.

## 6.4 Equipment Installation

**6.4.1** Flood protection measures shall be provided for the PV modules, DC combiner boxes and inverters. Measures such as setting up a flood bank or raising the lowest elevation of equipment may be taken considering site conditions. The top of the flood bank or the lowest elevation of the equipment shall not be lower than the standard flood level plus 0.5 m freeboard. When the site is likely affected by wind, wave or tide, wave run-up shall be considered.

**6.4.2** The building-mounted PV DC generation system should not be installed at a height within the reach of personnel, and electrification warning signs shall be set in a conspicuous position, and effective electrical safety measures shall be taken.

**6.4.3** For the building-attached PV DC generation system, the structural and electrical safety of buildings shall be evaluated, and shall meet relevant requirements.

## 6.5 DC Cable Layout

**6.5.1** The layout of DC cables shall comply with the current national standard GB 50217, *Code for Design of Cables of Electric Engineering* .

**6.5.2** DC cables exposed to the air shall be fixed and protected from direct sunlight.

**6.5.3** The cable tray laid outdoors shall meet the requirements for heat dissipation and drainage.

**6.5.4** Direct buried cables in the PV array area should be laid in the same trench.

**6.5.5** For the photovoltaic power station installed on the water, the DC cables shall be laid in the elevated cable tray, and the installation height of the cable tray shall not be lower than the standard flood level plus 0.5 m freeboard.

**6.5.6** For floating photovoltaic power stations, the length of DC cables shall consider water level fluctuation.

**6.5.7** For the photovoltaic power stations on the water, the incoming and outgoing cable trays of the DC combiner box shall not block the navigation channel.

**6.5.8** For DC cables running through the building envelope, the existing cable channels of the building shall be utilized. If new channel is needed, waterproof and fireproof measures shall be taken.

# 7 Overvoltage Protection and Earthing

**7.0.1** The overvoltage protection and earthing of DC generation system for PV power stations shall comply with the relevant national standards GB 50797, *Code for Design of Photovoltaic Power Station*; and GB/T 32512, *Technical Requirements for Protection of Photovoltaic Power Station Against Lightning*.

**7.0.2** Surge protection devices shall be installed on the DC side of DC combiner boxes and inverters. The selection of surge protection devices shall comply with the current standards of China GB/T 18802.31, *Low-Voltage Surge Protective Devices—Surge Protective Devices for Specific Application Including D.C.— Part 31: Requirements and Tests for SPDs in Photovoltaic Applications*; and NB/T 32004, *Technical Specification of PV Grid-Connected Inverter*.

**7.0.3** For building-mounted photovoltaic power stations, the lightning protection system shall utilize that of the building, and the lightning protection design shall be carried out according to the lightning protection grade of the building.

**7.0.4** The metal frame of PV modules shall be well earthed.

**7.0.5** The PV array shall be earthed in a continuous and reliable manner, with an earthing resistance of less than 4 Ω.

# 8 Protection and Monitoring

**8.0.1** When the same MPPT module is connected with more than two PV strings, the PV strings shall be provided with DC fuses or DC circuit breakers for over-current protection.

**8.0.2** Inverters shall be provided, at the DC input side, with DC overvoltage protection, DC reverse polarity protection, and DC input overload protection; and at the AC output side with AC open-phase protection, overvoltage/undervoltage protection, overfrequency/underfrequency protection, short-circuit protection, anti-discharge protection, anti-islanding protection, and low voltage ride-through protection.

**8.0.3** DC combiner boxes shall be designed with monitoring function and data acquisition module. The data acquisition module should collect the following information:

**1** DC bus voltage and current of each module string;

**2** Total DC output current and power;

**3** SPD status;

**4** Status of DC circuit breaker.

**8.0.4** Inverters shall be designed with monitoring function and data acquisition module. The data acquisition module should collect the following information:

**1** Bus voltage, current and power at DC side;

**2** Bus voltage, current and power of each incoming branch circuit at DC side;

**3** Three-phase voltage, frequency, current and power at AC side;

**4** Daily, monthly, annual and total cumulative energy output at AC side;

**5** Real-time conversion efficiency, power factor and internal temperature;

**6** Startup and shutdown time of the day;

**7** Inverter status and fault alarm information.

**8.0.5** DC combiner boxes and inverters shall be able to communicate data with the monitoring system.

**8.0.6** The data acquisition devices in DC combiner boxes and inverters should be self-powered.

# Explanation of Wording in This Code

**1** Words used for different degrees of strictness are explained as follows in order to mark the differences in executing the requirements in this code.

**1)** Words denoting a very strict or mandatory requirement:

"Must" is used for affirmation; "must not" for negation.

**2)** Words denoting a strict requirement under normal conditions:

"Shall" is used for affirmation; "shall not" for negation.

**3)** Words denoting a permission of a slight choice or an indication of the most suitable choice when conditions permit:

"Should" is used for affirmation; "should not" for negation.

**4)** "May" is used to express the option available, sometimes with the conditional permit.

**2** "Shall meet the requirements of..." or "Shall comply with..." is used in this code to indicate that it is necessary to comply with the requirements stipulated in other relative standards and codes.

# List of Quoted Standards

| | |
|---|---|
| GB/T 13539.6 | *Low-Voltage Fuses—Part 6: Supplementary Requirements for Fuse-Links for the Protection of Solar Photovoltaic Energy Systems* |
| GB/T 18802.31 | *Low-Voltage Surge Protective Devices—Surge Protective Devices for Specific Application Including D.C.—Part 31: Requirements and Tests for SPDs in Photovoltaic Applications* |
| GB/T 30427 | *Technical Requirements and Test Methods for Grid-Connected PV Inverters* |
| GB/T 32512 | *Technical Requirements for Protection of Photovoltaic Power Station Against Lightning* |
| GB/T 34581 | *General Technical Requirements for Photovoltaic System DC Circuit Breakers* |
| GB 50217 | *Code for Design of Cables of Electric Engineering* |
| GB 50797 | *Code for Design of Photovoltaic Power Station* |
| NB/T 32004 | *Technical Specification of PV Grid-Connected Inverter* |
| NB/T 42073 | *Electric Cables for Photovoltaic Systems* |